Environmental
Case Studies

Melissa R. Wuellner

Kendall Hunt
publishing company

Cover image © Shutterstock.com

publishing company

www.kendallhunt.com
Send all inquiries to:
4050 Westmark Drive
Dubuque, IA 52004-1840

Copyright © 2014 by Kendall Hunt Publishing Company

ISBN 978-1-4652-3797-2

All rights reserved. No part of this publication may be reproduced, stored in a retrieval system, or transmitted, in any form or by any means, electronic, mechanical, photocopying, recording, or otherwise, without the prior written permission of the copyright owner.

Printed in the United States of America
10 9 8 7 6 5 4 3 2 1

Dedication
To my teaching inspirations—Tom Lauer, Dave Willis, and Denise Peterson.
And to my life inspirations—Ryan and Lily.

contents

Themes

I. Sustainability, Conservation, and Preservation

    A. What do the Terms "*Sustainability*," "*Conservation*," and "*Preservation*" Mean? 1

II. World Population Growth

    A. Exploring Human Population Growth Policies Worldwide 3

III. Climate: Global and Local

    A. World Climate Scenarios 7

    B. What is Your Carbon Footprint? 11

IV. Sustainable Agriculture

    A. GMO Issues: Europe Versus the United States 13

    B. Cows or Corn? Considerations of Grassland-Cropland Conversion Decisions 16

V. Natural Resources Management

WATER

    A. Water Wars: The Battle for "Blue Gold" 19

    B. Too Much Water on My Land! Benefits and Concerns Related to Tile Drainage and Crop Fields 22

METALS

    C. Is Recycling Worth the Cost? A Case of Community Curbside Recycling Programs 25

OIL AND PETROLEUM

    D. To Frack or not to Frack? That is the Question...27

VI. Alternative Energy Development

ELECTRICITY

    A. Developing a Sustainable and Clean Energy Future with Tidal, Wind, and Solar Power: Exploring the Pros and Cons..29

PETROLEUM

    B. What is So Great About Ethanol? 31

VII. Terrestrial Ecosystems and Wildlife Issues

    A. Establishing Preserves to Aid Endangered Species: The Good and the Bad...................35

    B. The Dingo that Divides Us: Controversies of Reintroducing and Protecting Predators 38

VIII. Freshwater and Marine Ecosystems and Fisheries Issues

    A. Those Darn Dams! The Case of Gavins Point Dam, South Dakota-Nebraska 41

    B. Super Salmon! Should Genetically Engineered Salmon be Allowed on US Dinner Plates?.......45

INTRODUCTION

Contemporary environmental issues intersect science and human values and ethics. Concerns regarding the environment and its future are discussed daily in the media and in classrooms across the globe. The inaugural Earth Day in the US in 1970 was born out of air and water pollution problems brought to public attention by the media and books such as Rachel Carlson's *Silent Spring*. Since that time, the major environmental issues that garner attention have evolved as prior crises have been resolved or abated and new information comes to light. For example, topics such as climate change and protecting or reintroducing endangered species are now common.

New challenges will continue to arise as the world population continues to swell and the natural resources of our planet—both renewable and non-renewable ones—become increasingly limited. How can we sustain these life supporting resources in the light of such daunting challenges to ensure that future generations will be able to meet their needs? How do we conserve our wilderness areas or preserve biodiversity so that our great-great grandchildren and their great-great grandchildren enjoy nature as we see it now? Finding creative and long lasting solutions to such complex problems will require critical thinking about these issues and the impacts that individual actions, industrial practices, and government policies have on the environment at both the local and global scale. Deep engagement with scientific information and reconciling it with human needs, desires, and behaviors will be essential to being a responsible world citizen.

This collection of case studies was designed to prepare citizens for current and future environmental challenges through active learning. The cases provided herein are based on specific issues of the day or a composite of many specific examples. Very few cases have a single correct answer. Most require an in-depth analysis of alternative solutions, vetting pros and cons, weighing evidence, and exploring personal values and ethics to determine a course of action. In these cases, there is no right or wrong answer—just answers that are either well supported or not. Because of the nature of these cases, working through them may be slightly uncomfortable for students at times. Higher-order critical thinking skills will be required rather than rote memorizations and superficial application of concepts. But these cases will hopefully challenge preconceived notions, develop interest in today's environmental issues, and prepares informed citizens who consider how factors such as personal consumptive habits or other actions such as voting for particular political candidates can impact the environment.

This textbook is designed for introductory courses on environmental or natural resources conservation or similar subjects that are taught face-to-face, online, or any combination of the two. Students studying any major will benefit from these activities because working through these case studies can aid in developing the soft skills that employers in all fields of study require of today's college graduates (Crawford et al. 2011; NACE 2013). Each case challenges a student's problem solving skills. Effective communication (mostly written but some oral) and information literacy are strongly emphasized throughout the book, and many cases require teamwork. Refining these soft skills will help prepare students for the world of work after graduation.

Embrace the challenge of delving deeper into complex environmental issues and finding creative solutions to build a sustainable future. I hope this book stimulates you, the reader, to become a better environmental steward or to at least pause to think about the environment during your day-to-day activities. Good luck on your learning journey!

Melissa R. Wuellner
South Dakota State University

Literature Cited

Crawford, P., S. Lang, W. Fink, R. Dalton, and L. Fielitz. 2011. Comparative analysis of soft skills: What is important for new graduates? Accessed August 2013, http://www.aplu.org/document.doc?id=3414.

National Association of Colleges and Employers (NACE). 2013. Job outlook. Accessed August 2013, http://www.naceweb.org/Research/Job_Outlook/Job_Outlook.aspx?referal=research&menuID=69&nodetype=4.

section 1

Sustainability, Conservation, and Preservation

Case IA

What do the Terms "Sustainability", "Conservation", and "Preservation" Mean?

INTRODUCTION

The terms "sustainability," "conservation," and "preservation" are often used interchangeably when discussing environmental issues, but each term actually has a distinct meaning. In this assignment, you will explore the definitions for each of these terms, the relationships between these terms, and how each term can be applied to the management of natural resources within certain systems and your own life.

PART I: DEFINITIONS AND RELATIONSHIPS

a) Research the definitions for "sustainability," "conservation," and "preservation" using multiple reliable resources. Once you have completed your research, describe the definition of each term in your own words. Be sure to properly cite the resources you used to create your definitions. Your definitions should apply to the environment and natural resources in general, rather than a specific case or system.

b) In one paragraph, describe how these terms relate to one another. Provide examples that illustrate the specific relationships between each of these terms or create a conceptual model that describes these relationships. If you chose to do a model, the model should also be explained verbally. Any examples or conceptual models that were used from other sources should be properly cited.

PART II: APPLICATION

a) Choose one system: ecological, agricultural, or industrial. Conduct research using RELIABLE resources to find examples of practices that illustrate your definitions for "sustainability," "conservation," and "preservation." You should be able to find at least one specific example for each term within your chosen system. Describe this practice and how it illustrates the definition of that particular term in that system in one paragraph. Be sure to properly cite your research throughout your description.

b) Describe how your definitions can be applied to renewable and non-renewable resources in general (not just specific examples of those resources) in one paragraph. Before beginning your analysis, make sure that you understand the difference(s) between "renewable" and "non-renewable" resources. In your analysis, consider the following questions:

> Can non-renewable resources be considered "sustainable" resources?
>
> Can renewable resources be used or managed in an "unsustainable" manner?
>
> Should renewable resources be either "conserved" or "preserved"? Why?
>
> Should non-renewable resources be either "conserved" or "preserved"? Why?

You are welcome to cover other ideas in your analysis, but these are a few questions to get you started in your discussion.

c) How do you practice "sustainability," "conservation," and "preservation" in your own life as described in your definitions? In one paragraph, discuss at least one specific practice for each term and be sure to describe how this practice illustrates your definition of that term.

section 2

World Population Growth

Case IIA

Exploring Human Population Growth Policies Worldwide

INTRODUCTION

The world population currently exceeds seven *billion* people and is expected to reach nine billion people by the year 2050 (see past, current, and expected world population growth trends at http://www.census.gov/population/international/data/idb/worldpopgraph.php). Many scientists believe that we have already exceeded Earth's carrying capacity, which has caused considerable environmental damage due to ever-increasing demands for natural resources and accumulative levels of waste and pollution produced by human consumptive patterns and behaviors. Some pro-environmental groups believe that the ultimate solution to some of the world's most pressing environmental issues is to limit or control current and future population growth worldwide.

Per capita population growth rates in developing countries (i.e., less industrialized nations with a relatively lower gross domestic product) are often higher than those in developed countries (i.e., industrialized nations with a higher gross domestic product). Thus, efforts to stabilize or reduce population size often target developing counties rather than developed countries. Many developed countries financially support governmental policies designed to curb per capita population growth rates in developing countries either by providing aid directly to the governments of those countries or through monetary support of international agencies such as the United Nations Population Fund. Governments of countries may adopt *coercive* methods (i.e., those that impose a decrease in the number of children that families can produce such as forced abortions or adoptions of "excess" children) or *passive* methods (i.e., those that allow the family to voluntarily choose the number of children produced by providing access to family planning, contraception, and voluntary abortions) in their population control policies. However, implementation of any governmental policy that limits the number of children per family is not without controversy or unintended consequences.

In this case study, you will research different population growth policies worldwide and the positive and negative consequences of such policies, and evaluate your own view on whether or not such policies will be economically, socially, and environmentally sustainable long-term.

DIRECTIONS

a) Research the governmental policies for population control in one country that either currently practices or has previously practiced coercive methods of population control (i.e., China, India, Iran) and one country that practices passive methods of population control (i.e., Singapore, Taiwan, Thailand) using multiple reliable resources. Summarize your research in two paragraphs, one paragraph for each selected country. In your summary, describe the relative "success" of those policies, citing the changes in birthrates per capita compared to pre-policy rates.

b) Describe at least three positive and three negative long-term consequences of each country's population control policy. The consequences should describe economic, social, or environmental changes as a result of policy implementation. If it helps, organize your research for this question into a table similar to the one provided below before summarizing your research.

Consequences	Chosen Country A (Coercive Policies)	Chosen Country B (Passive Policies)
Positive	1) 2) 3)	1) 2) 3)
Negative	1) 2) 3)	1) 2) 3)

Write one paragraph for the positive consequences and one paragraph for the negative consequences for each country. This will result in a total of *four* paragraphs for this question. Note that some of the consequences may be similar between countries that implement coercive policies compared to those that implement voluntary policies, and other consequences will be different between the two countries.

c) What are the *ethical* consequences of any population control policy, whether coercive or passive? Describe at least three consequences in one paragraph.

d) What other factors besides government policies contribute to a decrease in birthrates per capita in a country? Describe at least three factors in one paragraph. Research the factors that have influenced birthrates in other countries that do not have population control policies currently in place using multiple reliable resources.

e) Based on your research, do you believe developed countries such as the United States, Canada, and England should financially support population control policies in developing countries? In your analysis, consider whether such policies are economically, socially, and environmentally sustainable in the long term and the consequences of implementing such policies as you described in parts b and c above.

Below are a few resources to help you start your research. However, you are highly encouraged to find other reliable resources of information.

Cincotta, R. P., and R. Engelman. 1997. Economics and Rapid Change: The Influence of Population Growth. Population Action International, Occasional Paper 3, Washington, D.C.

Findlay, A. M., and A. M. Findlay. 1987. Population and Development in the Third World. Routledge, New York.

Rust, D. L. 2010. "The Ethics of Controlling Population Growth in the Developing World." *Intersect* 3(1):69–78.

Smith, A. W. 1987. "Beyond Family Planning" ed. L. Grant. Pages 1–7 in *The Case for Fewer People: The NPG Forum Papers*. Santa Ana, California: Seven Locks Press.

United Nations Fund for Population Activities. 2012. By Choice, Not by Chance: Family Planning, Human Rights, and Development: State of the World Population 2012. New York: United Nations Fund for Population Activities.

section 3

Climate: Global and Local

Case IIIA

World Climate Scenarios

INTRODUCTION

Since the industrial revolutions of Europe in the 1700s and of North America in the 1800s, the emissions of greenhouse gases such as carbon dioxide (CO_2) have escalated, and scientific evidence strongly suggests that these emissions have led to changes in climate patterns across the globe. At the current and projected rates of greenhouse gas emissions, the expected CO_2 concentrations in Earth's atmosphere is expected to reach 976 parts per million (ppm) by the year 2100. This is expected to lead to a 5°C (41.0°F) increase in average air temperature above pre-industrial temperature averages. At this level, oceans are expected to rise at least 2 meters (6.6 feet), resulting in massive coastal land losses worldwide. In response to these alarming statistics, countries are working on international cooperative agreements to reduce greenhouse gas emissions on a global scale.

The Intergovernmental Panel on Climate Change (IPCC), established by the United Nations Environment Programme and the World Meteorological Organization, is the leading body that provides scientific analyses on climate change and its potential environmental, social, and political consequences (http://www.ipcc.ch/index.htm). Current IPCC goals are to maintain or reduce atmospheric CO_2 levels between 350 and 450 ppm between now and 2100. This will help to reduce increases in average air temperatures between 1.5 and 2°C (34.7–35.6°F) compared to pre-industrial air temperatures and result in far greater land loss due to rising ocean levels.

Current atmospheric CO_2 levels are influenced by three major factors: fossil fuel emissions, deforestation, and carbon sequestration. Fossil fuel emissions and deforestation add CO_2 to the atmosphere, whereas carbon sequestration removes CO_2 from the atmosphere. Different countries emit different levels of fossil fuels annually based on the level of economic industrialization and the consumptive habits of its citizens. For example, the poorest developing countries emit the lowest levels of CO_2 annually as the economy of those nations is largely based on agriculture and a low percentage of their citizens drive motorized vehicles. Conversely, the richest developed countries emit the highest levels of fossil fuels due to a largely industrialized economy and a high rate of motorized vehicle use amongst its citizens. However, many developing countries that are experiencing a shift to a more industrialized economy are beginning to reach or exceed fossil fuel emissions levels observed in developed countries.

Increased population growth is increasing demand for land and other natural resources produced by forests, thus promoting deforestation at a rate of approximately 7.3 million hectares worldwide per year, or roughly an area the size of Panama. Carbon dioxide is emitted as forests are converted for other land uses such as agriculture or urban and industrial development. Deforestation reduces the rate of carbon sequestration, because trees and other plants are no longer present to absorb atmospheric CO_2 during photosynthesis. Replanting forests (i.e., afforestation) can help reverse this trend.

In this case study, you will be simulating potential agreements to reduce carbon emissions and deforestation and to increase afforestation to examine the commitments that will be required of developed and developing countries to reduce global atmospheric CO_2 levels between now and 2100. The goal is to reach the IPCC recommendations (e.g., maintain or reduce atmospheric CO_2 levels between 350 and 450 ppm between now and 2100) to reduce potential temperature increases and coastal land losses worldwide. Once you have achieved these goals, you will reflect on what you learned during the course of the simulations.

PART I: INPUT LEVERS AND CLIMATE SCENARIOS

a) Visit the "Climate Interactive" website at http://climateinteractive.org/simulations/c-learn/simulation. Click on "Play as a Guest" and follow the instructions for accessing the simulator.

Once you log in, you should see the C-Learn Simulation tool and five separate graphs on the page. The three graphs on the left side of the page represent three input levers: "Fossil Fuel Emissions by Country Group"; "Emissions from Deforestation"; "and "Sequestration from Afforestation." You will be manipulating these input levers to produce the two graphs on the right side of the page: "CO_2 Concentrations in the Atmosphere" and "Temperature Change over Pre-industrial."

b) In the "Fossil Fuel Emissions by Country Group" graph, you will see three different lines. These lines represent the current and projected CO_2 emissions by three different country groups (e.g., Developed, Developing A, and Developing B). Definitions and examples of these country groups are as follows:

Developed—These are countries with historically industrialized economies and "medium-sized" populations. Currently, these countries emit the largest amounts of greenhouse gases per capita. Examples of these countries include the United States, Canada, England, France, Australia, New Zealand, Japan, and South Korea.

Developing A—The economies of these countries are currently shifting from agriculture to large-scale industrial economies. Populations of these countries are large, and the rates of greenhouse gas emissions per capita are growing rapidly. These rates may or have already exceeded those in Developed countries in some cases. Examples of these countries include China, Hong Kong, India, Pakistan, Brazil, South Africa, and Mexico.

Developing B—Some industrial production may occur in these countries, but the economies are largely agricultural based. Populations tend to be among the smallest of all country groups, and the rates of greenhouse gas emissions per capita are among the lowest in the world. Examples of these countries include Iran, Iraq, Costa Rica, Cuba, Botswana, Libya, Afghanistan, and North Korea.

The lines you see in the graphs at this moment represent the CO_2 levels if no steps are taken to reduce current growth in fossil fuel use or to reduce atmospheric CO_2 levels. These are referred to as the "Business as Usual (BAU) Standards." Note the patterns that are projected in fossil fuel emissions as well as CO_2 levels and temperature increases in the two graphs on the right side of the page if no steps are taken.

In the inputs below the graph, you can change the year in which each group of countries will commit to ceasing any increases in CO_2 emissions ("Stop Growth Year"), the year in which that group of countries will implement policies or steps to reduce CO_2 emissions from current levels ("Reduction Start Year"), and by what percent that group of countries will reduce CO_2 emissions on a yearly basis ("Percent Annual Reduction"). Manipulate these years and the percentage in your scenarios by typing information directly into the text boxes.

c) In the upper middle graph, you will see "Emissions from Deforestation." The values in this graph represent an index of worldwide future land use, where a value of "1" represents continued constant emissions from deforestation at the current rate of 5.4 Gigatons of carbon/year and "0" represents NO deforestation occurring worldwide by 2050. A value of "0.5" represents a 50% reduction of CO_2 emissions resulting from deforestation by 2050 and lasting until 2100. Slide the bar below this graph to change the levels of deforestation in your scenarios.

d) In the lower middle graph, you will see "Sequestration from Afforestation." The values in this graph represent an index of annual removal of CO_2 from the atmosphere by replanted forests. A value of "0" means that no CO_2 will be removed from the atmosphere, but a value of "1" removes 1.6 Gigatons of carbon/year. A value of "0.5" removes half of that amount, or 0.8 Gigatons of carbon/year. Slide the bar below this graph to change the levels of afforestation in your scenarios.

e) Alter the years, percentages, and indices in all three graphs described above and click the "Run" button when ready. Once you click this button, you will see the impacts of your proposed scenario on future CO_2 emissions and the consequential temperature change in the two graphs on the right. Remember, the IPCC goals are to maintain or reduce atmospheric CO_2 levels between 350 and 450 ppm between now and 2100. If you did not meet these goals in your scenario, go back to the three input graphs and change the years, percentages, and indices again and click "Run." Repeat this procedure as often as necessary to achieve the IPCC goals. It is recommended that you record your inputs and results after each scenario to help you track your progress. Some of this information will help you answer the reflection questions below. Note that you may require several tries before the goals are achieved. Once you have reached these goals, complete the reflection questions below.

PART II: REFLECTION

Now that you have reached the IPCC goals in your scenario, reflect on your experiences by answering the following questions:

a) Describe your experience with the scenarios.

   How many attempts did it take to achieve the desired goals?

   What were your final inputs in terms of emissions, deforestation, and sequestration?

   What were the final atmospheric CO_2 levels and temperature changes by 2100?

b) Which group of countries will need to commit to the most substantial changes in terms of fossil fuel emissions? Why?

c) What role did "Stop Growth Year" and "Reduction Start Year" play in your scenarios? Did you find that these inputs had to be substantially altered or slightly altered in order to achieve the desired goals? Which input had a more substantial impact on your results? Support your answers with evidence from your various scenarios.

d) Based on your scenarios, which of the following will have a greater impact on CO_2 levels and temperature changes in the future: reducing fossil fuel emissions, preventing deforestation, or increasing afforestation? Support your answer with evidence from your various scenarios.

e) In your opinion, how realistic are these commitments to reducing fossil fuel emissions, preventing deforestation, and increasing afforestation? Consider the years in which groups of countries will need to make such commitments, potential political or social challenges, and economic issues in your answer.

Case IIIB

What is Your Carbon Footprint?

INTRODUCTION

A "carbon footprint" is the amount of greenhouse gas emissions produced by an individual on an annual basis as a result of consumptive behaviors and various personal choices. In this assignment, you will explore your own carbon footprint and how your actions influence that footprint. You will also analyze the carbon footprints and carbon-producing behaviors of your fellow classmates, the average US citizen, and those of the world. Based on these various evaluations, you will come up with feasible goals for reducing your carbon footprint, implement those goals into your daily lives, and determine whether you have successfully reduced your carbon footprint after several weeks.

PART I: EXAMINING YOUR CARBON FOOTPRINT

a) Visit the Nature Conservancy's "Carbon Footprint Calculator" at http://www.nature.org/greenliving/carboncalculator/. This calculator will take you through several questions regarding your everyday activities and choices. Begin by selecting the number of people who are currently living in your household from the appropriate dropdown box, and then click "For Me Only." Be honest and answer every question the calculator asks under each tab (i.e., Home Energy, Driving and Flying, Food and Diet, and Recycling and Waste). Once you have completed all of the questions, click on the "Results" tab to view your total carbon footprint score as well as information regarding each of the categories to your overall score. You will also receive information on the average US carbon footprint and the average world carbon footprint.

b) In one paragraph, describe your carbon footprint by answering the following questions:

What were your total estimated emissions per year?

Which categories (home energy, travel, food and diet, recycling) had the greatest influence on your overall footprint and which had the least influence?

Were you surprised at these results? Why or why not? What personal choices or consumptive behaviors have influenced your footprint?

c) Compare and contrast your carbon footprint for your household to that of a fellow classmate. Your instructor may ask you to complete this as an in-class discussion or an out-of-class exercise. In your comparison and contrast, address the following questions:

Were your total emissions higher or lower than those of your classmate?

Which categories (home energy, travel, food and diet, recycling) were different and similar to your classmate's?

What personal choices or consumptive behaviors of your classmate lead to these patterns? How are these choices or behaviors similar or different from your own?

d) Compare and contrast your carbon footprint for your household to the carbon footprint of the average US citizen as well as the average world citizen. In your comparison and contrast, address the following questions:

Were your total emissions higher or lower than those of the United States and the world?

Which categories (home energy, travel, food and diet, recycling) were similar and substantially different compared to a US citizen and a world citizen?

What personal choices or consumptive behaviors of US and world citizens lead to these patterns? How are these choices or behaviors similar or different from your own?

PART II: REDUCING YOUR CARBON FOOTPRINT

a) Now that you have thoroughly analyzed your carbon footprint and the choices and behaviors that contribute to that footprint, describe three specific and measurable goals for reducing your carbon footprint that you can feasibly implement in the next three to six weeks. In describing each goal, address the following:

Describe how you plan to achieve each goal. In other words, what new behaviors or consumptive patterns will you adopt?

How will you determine whether each goal has been met? What measurements could you take? Make sure your measurement ties directly back to your goal. Envision the "end product" of your goal and how your progress can be tracked or measured as you work toward that goal.

Here is an example: if your goal is to reduce your footprint in the driving category by reducing the miles you drive per week by 20%, you might decide to bike to school or work three days per week. Once that behavior has been implemented, you could measure your progress by logging the number of driving miles in the subsequent weeks and then compare your driving miles during these weeks to the number of miles you logged in the weeks before you began biking.

b) After three to six weeks of working toward your three goals, review your progress on each goal. How successful were you in meeting those goals? In describing your relative success, consider the following questions:

Did you make a small amount of progress or was the goal achieved in its entirety?

What evidence of goal attainment can you provide following the description you provided in question IIa above?

Was it easy or difficult to meet your goals?

Were some goals harder than others?

What did the goals cost you in terms of time, money, or energy?

section 4

Sustainable Agriculture

Case IVA

GMO Issues: Europe Versus the United States

INTRODUCTION

Genetically modified organisms (GMOs) are organisms whose genetic material has been altered or "engineered" with the genetic material of another bacterial, viral, plant, or animal species to produce a desirable trait. In the agriculture sector, genetic engineering has produced GMO crops that are resistant to disease, pests, herbicides, or drought, are more flavorful, or have a longer shelf life.

Public acceptance of GMO crops for animal and food consumption varies worldwide. One of the striking contrasts between public and governmental support for GMO crops can be found in comparisons between Europe and the United States. Pro- and anti-GMO views can be found on both sides of the Atlantic Ocean, but generally anti-GMO sentiments are stronger in Europe. Overall, GMO supporters believe the development and cultivation of such crops allows farmers to stay competitive worldwide by increasing yields and decreasing potential losses in revenue. Increases in yields will also help feed a growing world population and provide more sustainable agriculture. Opponents of GMO have labeled these crops as "Frankenfoods" and believe the risks of such crops to human health and the environment are too high of a cost to pay for increased yields. Those who are anti-GMO argue that these crops are not sustainable given these costs.

Public opinion can sway government policies. Compared to the Unites States, the European Union (EU) more strictly regulates the use, importation, and sale of GMO crops for animal and human consumption. Scientific studies have examined the potential positive and negative economic, environmental, and human health effects of GMOs, and the information contained in those studies has been used and misused by both sides of the debate to further influence public opinion and governmental policies related to consumption and the need for labeling of food products that contain GMO crops.

In this case study, you will explore differences in public attitudes and governmental policies between the EU and the United States, review the science on the impacts of GMOs and the limitations of those studies, and explore your own opinions on GMOs and their sustainability.

DIRECTIONS

a) Review the current policies on consumption and labeling GMOs and foods produced from them for human and animal consumption in the EU and the United States. Summarize these policies in two paragraphs, one on the EU and the other on the United States.

b) Review and summarize scientific evidence on the potential economic (both to the individual consumer and to the industry as a whole), environmental, and human health benefits and risks of GMO crops using multiple reliable resources of information. It may help to organize your information into a table similar to the one below prior to writing your summary. Cite your sources of information appropriately.

	Benefits of GMOs	Risks of GMOs
Economic (Individual Consumer)	1) 2) 3)	1) 2) 3)
Economic (Industry)	1) 2) 3)	1) 2) 3)
Environmental	1) 2) 3)	1) 2) 3)
Human Health	1) 2) 3)	1) 2) 3)

c) What are some of the limitations of the scientific studies you summarized in question b above? Consider the following questions:

Do the results of the experiments or analyses in these studies accurately reflect potential positive or negative economic, environmental, or human health consequences in both the short-term and long-term?

Are the limitations of the study major or minor?

In your opinion, do the limitations of the study invalidate the conclusions of the study? Why or why not?

d) Are your personal views on GMO crops similar to those of the EU or the United States? Why? Describe your views in one paragraph.

e) Consider all of the economic, environmental, and human health consequences (good and bad) of GMO crop production and consumption. In your opinion, are GMO crops sustainable? Support your definition with the definition of SUSTAINABILITY used in this course as well as your research on question b above.

Below are a few resources to help you start your research. However, you are highly encouraged to find other reliable resources of information.

Godard, O. 1997. "Social Decision-making Under Scientific Controversy, Expertise, and the Precautionary Principle." eds. C. Joerges, K.-H. Ladeur and E. Vos. Pages 39–73 *in Integrating Scientific Expertise into Regulatory Decision Making—National Experiences and European Innovations.* Baden-Baden, Germany: Nomos Verlagsgesellschaft.

Lynch, D., and D. Vogel. 2001. The Regulation of GMOs in Europe and the United States: A Case-study of Contemporary European Regulatory Politics. New York: Council on Foreign Relations Press.

National Research Council. 2000. Genetically Modified Pest-protected Plants. Washington, D.C.: National Academies Press.

National Research Council. 2010. Impact of Genetically Engineered Crops on Farm Sustainability in the United States. Washington, D.C.: National Academies Press.

Case IVB

Cows or Corn? Considerations of Grassland-Cropland Conversion Decisions

INTRODUCTION

One of the most substantial land use changes in recent US history is the accelerated conversion of grasslands to cropland. These grasslands either were, or could be, used for grazing cattle or to restore prairies and the ecosystem goods and services they provide (Wright and Wimberly 2013). The majority of conversions are occurring in the western portion of the US Corn Belt, which includes Iowa, Kansas, Minnesota, Nebraska, and North and South Dakota. Agricultural producers and other landowners make the decision to convert land to row crop agriculture based on economic factors, government policies, social issues, personal values, and environmental constraints (Turner et al. 2013). Government policies and incentives in the United States such as the Farm Bill, ethanol subsidies, and crop insurance programs have enticed conversion by reducing some of the risks associated with row-crop agriculture and increasing profit margins. However, many advocacy organizations are working to convince farmers that cultivating grasslands rather than row crops can be profitable while also maintaining ecosystem integrity (see EcoSun Prairie Farms (2011) for one such example).

In this case study, you will advise someone faced with the decision of whether to continue preserving and conserving grasslands for cattle grazing or to convert the land to row-crop agriculture. You will evaluate economics, government policies, environmental consequences, social issues, and personal values in determining and justifying your recommendation.

SCENARIO

Your Grandpa Bob and Grandma Susie purchased a 200-acre farm in eastern South Dakota not long after they married in 1957. They took pride in the place and used the land for multiple purposes; 60% of the property was devoted to a corn-soybean-wheat rotation, 30% of the land was used for a small calf-cow operation (including pastures and stock ponds), and 10% of the land was either in shelterbelts or native grasslands to reduce land erosion (primary) and to provide some hunting and wildlife viewing opportunities (secondary).

Your dad was the only child that Grandpa Bob and Grandma Susie had. Your dad decided that he was more passionate about a career in aviation than in farming and moved out of state permanently after completing college. But he would often return to the farm to help out your grandma and grandpa with various day-to-day operations.

In 1993, Grandpa Bob and Grandma Susie decided they were getting "too old" to farm anymore, but neither had any desire to move from the land they loved so much. Grandpa Bob enrolled his acres of cropland into a 10-year Conservation Reserve Program (CRP) contract which paid him an annual rental rate to convert the land back to grasses. He re-enrolled his acres again in 2003, but the contract expired in 2013. Also back in 1993, Grandpa and Grandma's younger neighbor—a rancher—asked whether he could lease annual grazing rights on the 30% of the land that Grandpa used to run his calf-cow operation and Grandpa agreed. After you were born, your dad would often take you and your mom back to the farm, and you and the entire family shared many great memories of hunting deer, waterfowl, and pheasants and listening to beautiful songbirds on the farm.

When Grandpa and Grandma recently passed on, your dad inherited the farm. He had no plans to move back onto the land, but he still wanted to keep the property so he could go back whenever he wanted as he used to do when his parents were alive. Recently, the ranching neighbor who had been grazing his cattle for the past 20 years on your grandpa's property called your dad to ask if he could continue grazing and expand his grazing lease to include 60% of the total property as his operation was growing. A new neighbor, who farms several thousand acres of corn and soybeans in the area, is also interested in converting 60% of Grandpa's farm to row crops and would pay your dad to lease the land.

Your dad knows that in order to continue paying taxes on and to maintain the beloved farm, he will need to lease the land to one of these producers. But, he's torn. On the one hand, your dad knows that he has the potential to make more money on his lease to the farmer than the rancher due to current corn and soybean prices, government incentives and policies, and reduced risk of crop loss with genetically modified varieties and crop insurance programs. On the other hand, your dad has known the rancher for a long time and believes (based on his own research) that grazing does a better job at preserving the ecological integrity of the landscape that supports all of the wildlife and recreational opportunities the family has shared over the years. Since you will one day inherit the land and are currently receiving a quality college education in everything from economics to ecology, your dad looks to you for advice. Should he lease the land to the farmer or the rancher?

DIRECTIONS

1) Describe the pros and cons of converting grasslands on Grandpa's farm to row crops or expanding grazing operations in terms of potential economic gains and risks, government support, environmental impacts, social issues, and personal values. Some of the pros and cons (i.e., economic, environmental, and social impacts) may require research from reliable resources, but others (i.e., personal values) are based on your own views. You may find it helpful to organize your thoughts by using a table similar to the one below:

	Converting Grasslands to Croplands		Expanding Grazing Operations on Grasslands	
	Pros	Cons	Pros	Cons
Personal economic gains/risks				
Government support				
Social issues				
Environmental impacts				
Personal values				

2) Since you are currently in college and can't talk to your dad in person, compose an e-mail to him recommending whether he should lease the land to the farmer or the rancher. Support your recommendation with your research and views from question #1 above. Make sure your argument is persuasive and convinces your dad that this is the right option for you and your family.

Literature Cited:

EcoSun Prairie Farms. 2011. Grass Roots: A Prairie Farm story. Accessed March 2013, http://thegrassrootsfilm.com/index.html.

Turner, B. L. R. Gates, T. Nichols, M. Wuellner, B. H. Dunn, L. O. Tedeschi. 2013. "An Investigation into Land Use Changes and Consequences in the Northern Great Plains Using Systems Thinking and Dynamics." *Proceedings of the 31st International Conference of the System Dynamics Society.*

Wright, C. K., and M. C. Wimberly. 2013. "Recent land use change in the western Corn Belt threatens grasslands and wetlands." *Proceedings of the National Academy of Sciences* 110(10):4134–4139.

section 5

Natural Resources Management

Case VA

Water Wars: The Battle for "Blue Gold"

INTRODUCTION

Over the history of mankind, civil and international conflicts have been waged over the use and allocation of natural resources. In more recent decades, the most prominent battles have dealt with oil, but many historical wars and skirmishes have been waged over basic necessities such as water. Rivers know no political boundaries—they flow between states, territories, and nations. As the world population continues to grow and more acres of land undergo desertification for a variety of reasons, water will likely become increasingly limited for human use and consumption (Carius et al. 2004). Many believe the next major civil and international wars will be fought over water or "blue gold" (Carius et al. 2004; Quinn 2012; Michel 2013). This fictitious case study is a glimpse into the potential future and is based on several past or current water use and allocation issues throughout the world, including the United States.

BACKGROUND

The fictitious Aqua River is a 2,500-km river that flows freely (i.e., no dams) between seven city-states in one country (Bajeer—not a real country) before flowing into a second country (Cela—again, not a real country) and then meeting the ocean. Most (83%) of the river's kilometers are located in Bajeer. The Aqua River is primarily a snowmelt-driven system, meaning that springtime snowmelt in the four mountainous upper city-states of Bajeer provide most of the river's year-round flow. Very little precipitation in the form of rain falls in the lower Aqua River basin, which consists of the three city-states and Cela. In fact, most of the lower basin is classified as desert.

Demand for water has been increasing in the last 20 years throughout the Aqua River basin for various reasons. The population of Bajeer as a whole has increased 15%; certain city-states, including two in the upper basin and one in the lower basin have experienced the largest "booms" in population growth due to immigration. In the lower basin in both Bajeer and in Cela, unprecedented land conversion from native vegetation to row crop agriculture has increased irrigation demands from the river. In fact, 70% of the water use in the lower Aqua River basin is for agricultural purposes.

In addition to these increases in water demand by humans throughout the entire basin, water supplies are threatened. In Cela, non-native plants have invaded the riparian corridors along the river. These plants are "thirstier" than native plants and have effectively lowered groundwater levels and contributed to the desiccation of wetlands. The non-native plants are threatening to invade upstream into Bajeer. According to scientists in Bajeer and Cela, climate change is expected to increase average ambient temperatures in the Aqua River basin by 3°C by 2050. Such an increase is expected to contribute to a reduction in snowpack in the upper four city-states of Bajeer, virtually eliminating all rain in the lower half of the basin, and increasing rates of evapotranspiration throughout the entire basin. It is expected that instances of drought will increase and these episodes will last for longer periods than observed in the past. Consequently, scientific models predict that water demand will exceed water supply in at least eight of the next 20 years.

The year is now 2026. Average ambient temperatures throughout the Aqua River basin have risen 1.2°C and snowpack in the upper basin has declined 17% below average levels during the last seven years. Invasive plants have wiped out native vegetation in the lowest third of the Aqua River basin and are quickly marching upstream. Alarmed by these facts, the four upper city-states of Bajeer are looking for solutions to keep the upper basin water within their boundaries and under their control. The governments of these four city-states believe the lower three city-states and Cela have done a poor job in managing water for agricultural use, controlling invasive plant species, and allocating water to growing population centers. The four city-states are working on a major dam project that would reduce downstream flows by 30%. In some years, this reduction in flows means that the Aqua River would not reach the ocean.

Upon hearing this, the governments of the three lower city-states of Bajeer and Cela were outraged, believing that the upper basin governments will be stealing Aqua River water from their downstream neighbors if the dam is built. They also believe that the upper city-states have done very little to regulate water for growing populations in the upper basin. These lower basin governments have vowed that if the dam project is completed, they will declare war on the upper basin, sparking a civil and international conflict that has the potential to draw in other countries in this volatile region of the world. The United Nations is visiting the region to negotiate peace. In order to satisfy all parties in this conflict, the governments of the seven city-states of Bajeer and the government of Cela must agree to acceptable means for conserving water and augmenting water supplies.

Pre-class Preparation:

Prior to class, spend time researching real-life "water wars" and how they were, or potentially could be, resolved through water conservation and augmentation. Find established and creative solutions. List at least three ideas to conserve water and three ideas to augment water supplies.

In-class Activity:

Your instructor, who will be playing the role of the President of the United Nations General Assembly, will assign you to one of three groups: the government representatives of the upper four city-states of Bajeer; the government representatives of the lower three city-states of Bajeer; and the government representatives of Cela. Meet with your group to discuss a proposal for peaceful conflict resolution for the Aqua River. Use your pre-class research and the background of the case study to determine what your group demands from the other two groups in terms of water conservation and/or augmentation and what your group would be willing to do in return if your mandates are met. Be specific on your requirements and concessions; make sure the goals are measurable. For example, you could say, "We expect group X to augment water supplies by 18% within the next five years by doing. . . ." Write these demands and concessions in one paragraph from the entire group to give to the President of the United Nations General Assembly.

3) The President will review the three proposals and write a bulleted summary of the demands and concessions from each group on the board. Review the bulleted list. Where do the three proposals agree and disagree? Jot down notes on these areas.

4) Discuss potential areas of negotiation within your assigned group. What would your group be willing to do to meet the demands of the other two groups? What concessions do the other two groups need to make in return? Discuss and write about these ideas for the next 10–20 minutes. Elect one group spokesperson to present these ideas orally to the entire class. The President will amend the previously bulleted list during the presentations.

5) Are the three groups closer to an agreement? If complete agreement has been reached, then negotiations can end and peace has been achieved. If not, repeat the negotiation process until peace has been brokered for all three sides in this conflict.

Post-class Reflection:

1) In your opinion, was this activity "realistic"? Why or why not? Support your answer with specific examples from your own research on "water wars" throughout the world.

2) Do you believe you will witness a "water war" in any part of the world during your lifetime? Why or why not? Explain.

Literature Cited:

Carius, A., G. D. Dabelko, and A. T. Wolf. 2006. Water, Conflict, and Cooperation. Policy Brief: The United Nations and Environmental Security. Accessed July 2013, http://www.unep.org/disastersandconflicts/Portals/155/disastersandconflicts/docs/ecp/ecspr10_unf-caribelko.pdf.

Michel, D. 2013. Egypt, Ethiopia Water Dispute Threatens Nations. International Business Times. Accessed July 2013, http://www.ibtimes.com/egypt-ethiopia-water-dispute-threatens-nations-1324189.

Quinn, A. 2012. U.S. Intelligence Sees Global Water Conflict Risks Rising. Reuters. Accessed July 2013, http://www.reuters.com/article/2012/03/22/us-climate-water-idUSBRE82L0PR20120322.

Case VB

Too Much Water on My Land! Benefits and Concerns Related to Tile Drainage and Crop Fields

INTRODUCTION

Subsurface agricultural land drainage has been used widely in the United States since the early 1900s to remove excess water from soils, prevent crop damage or loss due to floods, and to help optimize the timing of spring plantings (Donnan 1976). In recent years, the rate at which subsurface drainage, also called tile drainage, has been installed in the United States has accelerated annually due to higher-than-average corn and soybean prices, increased rates of land conversion to row crops, government policies and incentives, and higher-than-average precipitation patterns (Lien and Orrick 2012). For example, the Bois de Sioux Watershed District in Minnesota approved permits for only 2.9 miles of subsurface tile in 1999. Ten years later, it permitted nearly 780 miles of tiling. In 2011, that number accelerated to nearly 1,560 miles of approved tiling (Lien and Orrick 2012).

Tile drainage has contributed to increased yields, reduced soil erosion rates, and improvements in groundwater quality as measured by a reduction in phosphorous (Fraser and Fleming 2001). However, the subsurface water that is removed from agricultural fields flows downstream to streams and rivers in the watershed. The increase in water volume results in more frequent and intense flooding for those living downstream, increased erosion of stream banks, and reduced water quality in terms of nitrogen concentrations and contamination by *Escherichia coli* bacteria (Lien and Orrick 2012).

The decision by farmers of whether or not to install tile drainage on his/her land is largely influenced by economics (e.g., can the cost of installation be recouped in the form of increased crop yields). But peer-pressure or persuasive arguments from various stakeholder groups may also sway decisions. Major stakeholders on each side of the issue may include:

Pro-tiling Groups	Anti-tiling Groups
Farmers in upstream communities	The local municipal water treatment plant
The local Farm Bureau	General environmental advocacy groups
Drain tile installation companies	Wildlife conservation advocacy groups

Persuasion from these stakeholders may come in a variety of forms, from an informal conversation at the local coffee shop to "Letters to the Editor" published in the community newspaper. In this case study, you will explore the pros and cons of tiling, assume the role of one of these stakeholder groups and argue your position to a local farmer who may be contemplating installing drain tiling on his/her corn and soybean fields.

EXERCISE

Your instructor may elect to complete this assignment as either an in-class "coffee shop" discussion and reflection activity or an individual "Letter to the Editor" writing and reflection assignment. Directions for both activities are provided.

IN-CLASS "COFFEE SHOP" DISCUSSION AND REFLECTION DIRECTIONS:

1) Spend time prior to class researching the various economic and environmental pros and cons of tiling using RELIABLE resources. You may find it helpful to organize your research by using a table similar to the one below:

Impacts	Pros	Cons
Personal economic gains/costs		
Downstream community impacts		
Soil health		
Water Quality		
Wildlife		
Flooding frequency and intensity		
Others		

2) In class, your instructor will assign you to one of the stakeholder groups. Meet with your fellow group members to discuss the pros and cons that each of you wrote individually and compile a master list of pros and cons of tiling.

3) As a group, use your list of pros and cons to craft a casual but convincing monologue that someone who represents your stakeholder group might say to a farmer who is contemplating installing drain tiles on his/her corn and soybean fields and whom you have happened to have run into at the local coffee shop. You will have three to five minutes to talk to this farmer who will be played by your instructor.

4) Listen to the monologues of the other stakeholder groups and jot down any additional pros and cons that your group did not identify in question #2 above. Either at the end of class or before the next class period, write a one- to two-paragraph response to the following question: If you were a farmer, would you choose to tile your land or not? Support your answer with the pros and cons derived from the in-class group discussions and monologues.

OUT-OF-CLASS "LETTER TO THE EDITOR" DIRECTIONS:

1) Research the various economic and environmental pros and cons of tiling using reliable resources. You may find it helpful to organize your research by using a table similar to the one below:

Impacts	Pros	Cons
Personal economic gains/costs		
Downstream community impacts		
Soil health		
Water quality		
Wildlife		
Flooding frequency and intensity		
Others		

2) Your instructor will assign you to one of the stakeholder groups. Write a "Letter to the Editor" of your community newspaper to convince local farmers in your area to install or refrain from tiling their land (depending on your group's position). Support your stakeholder's position with the research you described for question #1 above. Remember, your letter should be concise but impactful. Keep your letter no longer than two double-spaced pages

3) Your instructor will make all of the letters written by your classmates anonymous but available for the entire class to read either in hard copy form, via e-mail, or on your course website. Read all of the other letters from the other students. Jot down any additional pros and cons that you did not identify in question #1 above. Once you have reviewed the other letters, write a one or two paragraph response to the following question: If you were a farmer, would you choose to tile your land or not? Support your answer with the pros and cons derived from your own research and the letters of other students.

Literature cited:

Donnan, W. W. 1976. "An Overview of Drainage Worldwide." ed. American Society of Agricultural Engineers. Pages 6–9 in Proceedings from the Third National Drainage Symposium, St. Joseph, Michigan.

Fraser, H., and R. Fleming. 2001. Environmental Benefits of Tile drainage: Literature Review. Guelph, Ontario: University of Guelph.

Lien, D., and D. Orrick. 2012. Minnesota Farm Drain Tiling: Better Crops But at What Cost? Pioneer Press. Accessed March 2013, http://www.twincities.com/outdoors/ci_21445585 /minnesota-farm-drain-tiling-better-crops-but-at .

Case VC

Is Recycling Worth the Cost? A Case of Community Curbside Recycling Programs

INTRODUCTION

Approximately 10,000 towns in the United States support curbside recycling programs. Recycling is often one of first practices people think of or participate in to benefit the environment. However, even recycling isn't without some controversy as to whether it is economically feasible and advances conservation efforts. In this case study, you will explore some of the criticisms as well as additional pros and cons of recycling to determine whether your local community should support or eliminate a curbside recycling program.

SCENARIO

One year ago, you were elected to the city council of your hometown. The town's population is around 20,000 and the budget of the town includes support of services and amenities such as local utilities (electricity, water, gas), safety services (ambulance, police, fire), the city hospital, public pools and parks, as well as a weekly garbage pickup and curbside recycling services. Recent economic events, both locally and nationally, have incited concern about the town's budget. Revenues from local taxes are projected to decline over the next several years as a loss of job opportunities in the town may force citizens to relocate to other cities. The nine members of the city council—including you—have begun discussing ways to reduce the town's budget, which may include eliminating some of the services or attractions described above.

In the thick of discussions, your child returns home from college for a holiday break. During dinner one evening with the family, your son/daughter describes an interesting piece that his/her economics professor required the class to read called "Eight Great Myths of Recycling" by Daniel Benjamin (2003). You are intrigued by the article. Knowing that your town has a curbside recycling program that is subject to possible elimination from the town's budget, you ask your son/daughter for a copy of the article. You hope to discuss the reading at the next city council meeting.

DIRECTIONS

Critically read Benjamin's piece entitled "Eight Great Myths of Recycling." You can find a PDF copy of this article online or your instructor may provide a copy to the class. As you read, make note of areas where you may need to find more information to answer the questions below. You will discuss your answers to the questions below in class or in an online discussion forum.

a) In one paragraph, describe three points the author makes on which you agree. In a separate paragraph, describe three points the author makes on which you disagree. In both paragraphs, describe why you agree or disagree with each point by citing information from RELIABLE resources.

b) What are some additional pros and cons of recycling or using products made from recycled materials does the author not describe in his piece? Describe the pros and cons in two separate paragraphs and cite information from RELIABLE resources.

c) Based on your analysis of the piece in the two questions above, what would you recommend to your fellow city council members regarding the fate of the town's curbside recycling program? Should the program remain in the town's budget or be cut (i.e., eliminate curbside recycling in the town entirely)? Support your decision with your research.

Post-class Discussion Reflection

a) As the class discusses the responses to the three questions above, make note of any additional pros and cons of recycling or other points of the article with which students agreed and disagreed. Based on this discussion, would your prior recommendations remain as is or would they change? Why or why not? Describe in one paragraph.

Literature Cited:

Benjamin, D. 2003. Eight Great Myths of Recycling. Property and Environmental Research Center. Accessed June 2013 from http://perc.org/sites/default/files/ps28.pdf.

Case VD

To Frack or not to Frack? That is the Question.

INTRODUCTION

Hydraulic fracturing—commonly known as "fracking"—is the process by which water is sent miles below the Earth's crust to create fractures in rocky layers that release natural gas and petroleum. In recent years, the fracking industry has witnessed booms in states such as Pennsylvania and North Dakota. As of 2012, fracking has created more than one million jobs in the United States and 2.5 million jobs worldwide (King 2012). What are the benefits of fracking from an economic and social standpoint? What are the environmental, social, health, and economic consequences of fracking? Why are some communities supporting the development of a fracking industry in their area while other communities, and even entire countries, are banning fracking development? In this case study you will examine the pros and cons of hydraulic fracturing and the development of such an industry using role playing.

SCENARIO

You live in a rural community of 1,500 people. Beneath the town, there is a large geologic formation that has been proven to contain oil. This formation is so large in fact that it expands into two neighboring states. Geologic surveys estimate that four to five billion barrels of oil could potentially be recovered from the entire formation using current hydraulic fracturing technology. Several communities in the two neighboring states have begun attracting oil companies to develop oil wells in their area by selling mineral rights, providing monetary incentives, and creating hotels, restaurants, and other amenities to attract new out-of-state workers. Your town is feeling the pressure to do the same. Several oil companies have contacted the town council about initiating a fracking industry in and around the community. Certainly the new industry would bring in a new source of revenue to local businesses and the town budget. But are the potential gains worth the costs of a rapidly developing industry?

In order to determine whether the town should support the fracking industry, the town council is seeking input from its citizens. The town is divided: some are "pro-fracking" while others are "anti-fracking." The town council has invited members of the community to speak to both sides of the issue during the next public council meeting. The following individuals have been invited to speak:

Pro-Fracking

    Dale Williams—a local restaurant and hotel owner seeking to expand his business to support his growing family

    Mary Johnson—a real estate agent who also owns several rental properties in the town

    Merle Davidson—a lifelong resident (60+ years) of the town who wishes the United States to gain independence from "foreign oil"

Anti-Fracking

> Jessup Monroe—a local hunting and fishing guide concerned about the environmental impacts of fracking
>
> Dr. Hailey Bright—community family practice physician concerned about the potential health issues and physical safety of the oil workers
>
> Ann Riley—a stay-at-home mom and president of the Parent Teacher Association (PTA) concerned about public safety issues that may arise during the fracking "boom"

DIRECTIONS

In-class Town Council Meeting

Your instructor will assign you to one of the six roles described above. Research the benefits or concerns of fracking as they relate to your assigned character. To aid you in your research, look for information from other fracking "booms" and "busts" in other US states and why communities and countries are either "pro-fracking" or "anti-fracking." Use RELIABLE resources to prepare a three- to five-minute speech for your character to convince the town council to either support a growing fracking industry or to ban fracking development in your area. Present your speech to the class.

Post-class Reflection

As you listen to others' speeches, jot down any new information that either supports fracking or discusses the negative consequences of fracking. After class, take time to read your notes and complete any follow-up research on the pros and cons of fracking using RELIABLE resources. In one to two paragraphs, answer the following question: If you were a member of the town council, would you vote to support or ban fracking development in your community? Support your decision with what you learned in the class speeches and your own research.

Literature Cited:

King, G. E. 2012. Hydraulic Fracturing 101: What Every Representative, Environmentalist, Regulator, Reporter, Investor, University Researcher, Neighbor and Engineer Should Know About Estimating Frac Risk and Improving Frac Performance in Unconventional Gas and Oil Wells. Society of Petroleum Engineers, Paper 152596.

section 6

Alternative Energy Development

Case VIA

Developing a Sustainable and Clean Energy Future with Tidal, Wind, and Solar Power: Exploring the Pros and Cons

INTRODUCTION

As the world population increases, demand for electricity to power homes and businesses will also increase. Most electricity is currently generated using non-renewable resources that also emit many greenhouse gases into the atmosphere. One initiative to create a sustainable and clean energy future for the planet includes using renewable energy systems to generate electricity. Several commercially available renewable energy systems currently provide electricity to homes and businesses, including tidal, wind, and solar energy. But which energy system is best in creating a sustainable and clean energy future across the globe? In this assignment, you will critically evaluate the current production methods of wind, tidal, and solar energy and their potential contributions to our energy future.

DIRECTIONS

a) Research how tidal, wind, and solar energy systems work to generate electricity for homes and businesses using multiple RELIABLE resources. Briefly summarize your research in one paragraph for each energy system. Include information on whether the energy can be stored and, if so, how it is stored.

b) Describe at least three unique pros and three unique cons of each energy system. Some questions that may be considered in researching and evaluating the pros and cons may include (but are not limited to) the following:

What are the costs of the system to the producer and the consumer of electricity from that particular system?

Is that particular energy system available in all geographic areas of the globe?

How frequently will the energy system require routine maintenance?

Is energy storage for that system an issue or not?

Are non-renewable resources required to create the energy system and produce electricity?

Does the energy system or it's delivery to the consumer emit any greenhouse gases? If so, how much?

What are other potential impacts of that system to the environment, including living and non-living things?

You are welcome to cover other pros and cons not listed above, but these are a few questions to help get your research started. Describe three pros for a single energy system in a single paragraph before describing three cons in a second paragraph. Repeat the process for the other two energy systems.

c) In one paragraph, describe which energy system you believe is the most SUSTAINABLE. Support your choice with the definition of "sustainability" that is used in this course as well as your own research on the pros and cons of these energy systems.

d) In one paragraph, describe which energy system you believe will be the most likely to reduce the greatest amount of greenhouse gases in the atmosphere worldwide. Justify your choice with your research on the pros and cons of each energy system and any additional research related to greenhouse gas emissions and electricity generation.

Case VIB

What is So Great About Ethanol?

INTRODUCTION

Governments worldwide are incentivizing the development and production of clean, sustainable energy for motorized vehicles that reduces dependence on imported petroleum. For example, the European Union mandates that 10% of the entire demand for fuel among all member countries should be met through the use of biofuels. In the United States, the Environmental Protection Agency's goal is to increase the country's use of biofuels to 36 billion gallons by 2022. Ethanol is one of many types of biofuel but is the one most heavily promoted in various government policies across the globe. Most ethanol produced in the United States is corn-based but cellulosic biomass (e.g., sugarcane, switchgrass, corn stover, wood waste) may also be used. Critics and supporters of ethanol have debated government mandates that support the development and production of these biofuels. On the one side, ethanol does burn "cleaner" than petroleum and can help to build or revitalize rural economies. The plants that are grown and harvested for ethanol production can also absorb CO_2 from the atmosphere (i.e., carbon sequestration), thus potentially impacting global climate change. On the other hand, CO_2 inputs are often required to grow, tend, harvest, and transport those plant materials to be made into ethanol. Further, the promotion of ethanol has been blamed for the conversion of millions of acres of grasslands to croplands (to the detriment of wildlife and other ecosystem goods and services) and increased costs to the American public in the form of taxes and food prices.

In this assignment, you will explore the various social, economic, and environmental impacts of both corn-based and cellulosic ethanol. Based on your research and your personal experience, you will decide whether you believe either form of ethanol will result in a net reduction of CO_2 in our atmosphere and whether either form is socially, economically, or environmentally sustainable.

PART I: RESEARCH

In the table below, you will note several categories that will help you explore the pros and cons of both corn-based and cellulosic ethanol. Research each of these categories to help you understand both the pro-ethanol and the anti-ethanol sides of the debate. In certain categories, finding specific numbers or statistics is encouraged. However, if these do not exist, you should be able to describe the differences between corn-based and cellulosic ethanol in relative terms. You may find that for some categories, the information for both corn-based and cellulosic ethanol is the same.

Categories	Corn-based Ethanol	Cellulosic Ethanol
Greenhouse Gases – Inputs and Outputs		
Carbon sequestration		
Use of fossil fuel-based fertilizers		
Use of fossil fuels during harvest and transportation		
Carbon emissions during use in motorized vehicles		

Categories	Corn-based Ethanol	Cellulosic Ethanol
Greenhouse Gases – Inputs and Outputs		
Vehicle fuel efficiency (miles per gallon)		
Frequency of fueling relative to non-blended gases		
Direct Costs to Consumer		
Price at the pump		
Total cost of fuel/year (include frequency of fueling in calculation)		
Price of an E-85 vehicle compared to a "traditional" vehicle		
Other Social, Economic or Environmental Impacts		
Creation or loss of jobs (some sectors will gain and others may lose)		
Rural economic development impacts		
Changes in land conversion rates		
Impacts to wildlife		
Impacts to other ecosystem goods and services		
Taxpayer costs of subsidies		
Impacts on food costs		
Energy security and independence		

PART II: EVALUATION

Now that you have researched corn-based and cellulosic ethanol and the consequences (both good and bad) of their development and use, answer the following questions to evaluate your level of support for ethanol in general and for corn-based or cellulosic ethanol specifically.

a) Will the use of ethanol in general result in a net reduction of CO_2 emissions? If yes, will one form of ethanol (i.e., corn-based or cellulosic) have a greater impact than the other or will the impacts be similar? Support your answers with your research above.

b) Consider the definition of "sustainability" used in this course. Given this definition, answer the following questions:

    Do you believe ethanol use in either form is socially, economically, and environmentally sustainable? Why or why not? Support your answer with your research above.

    If you answer yes, describe whether you believe corn-based or cellulosic ethanol is more socially, economically, and environmentally sustainable. Support your answer with your research above.

c) Given your research and your own personal experience prior to this course, do you support ethanol from any source as an additive to gasoline for motorized vehicles? Support your opinion with your research and experience, addressing both the pro-ethanol and the anti-ethanol sides of the issue. If you support ethanol use, would you prefer corn-based or cellulosic ethanol use? Support your answer with your research from above.

section 7

Terrestrial Ecosystems and Wildlife Issues

Case VIIA

Establishing Preserves to Aid Endangered Species: The Good and the Bad

INTRODUCTION

This case study is not based on any single example of conflicts related to the establishment of wildlife protection areas or preserves. Rather, several international examples inspired this activity. As you work on this case study, you are strongly encouraged to research examples of real life human-wildlife conflicts in and around wildlife protection areas and preserves worldwide.

BACKGROUND

The fictitious developing nation of Oingo Boingo is home to a wealth of biodiversity, including many unique and rare species. One such species that has caught the attention of the world is the charismatic feline predator, the silver leopard (not a real species). This elusive creature occupies the mountainous region of northwestern Oingo Boingo and was recently featured on a highly popular nature documentary produced by the British Broadcasting Corporation (BBC). As noted in the television program, the silver leopard is at dire risk of becoming extinct within the next 30 years. The silver leopard was traditionally hunted for its luminous pelt. Despite being protected from hunting by law, poaching still continues (though at a much lower level). Human alterations of habitat and the hunting of the silver leopards' preferred prey items by local tribes also pose threats to the survival of this species. It is estimated that only 40 silver leopards are left in Oingo Boingo.

In response to a worldwide outcry to save the silver leopard from extinction, international conservation organizations such as the Save Our Wildlife Federation (not a real organization) have been receiving millions of dollars in donations to help establish a 2,500 km² wildlife preserve that would be akin to a US National Park, like Yellowstone or Yosemite. Once enough money has been raised, the organization plans to use the funds to offer the government of Oingo Boingo a monetary gift to be used to establish and maintain the preserve. Save Our Wildlife Federation believes the preserve would not only benefit the silver leopard but also many other endemic species of mammals and birds that occupy the same ecosystem.

Further support for the establishment of a wildlife preserve has been given by those who work in the ecotourism industry in the countries surrounding Oingo Boingo. Oingo Boingo currently has no ecotourism industry, but the BBC feature has indirectly created an interest by international tourists to visit the country in the hopes of seeing the rare cat and other unique fauna. Establishment of an ecotourism industry in Oingo Boingo could be a boost to the economy of this developing nation. Representatives from the tourism industry cite examples from other countries whose economies have benefited from wildlife based tourism. For example, wildlife based tourism accounted for 25% of Kenya's Gross Domestic Product (GDP) and more than 10% of the nation's employment rate in 2006 (see Okech 2010).

However, the potential establishment of the wildlife preserve has been met with opposition from many of the Oingo Boingo people. Creation of the preserve may affect at least three Oingo Boingo tribes in several ways. First, these tribes hunt some of the native ungulates that are found within and around the proposed boundaries of the preserve for subsistence. Establishment of the preserve would prohibit hunting of any species within its boundaries. Secondly, the tribes contend the preserve would eliminate access to some (not all) grazing and crop lands, which provide additional food sources to the tribe. Third, the tribe leaders believe grazing domestic livestock near the preserve's boundaries would put humans and livestock at risk of predation by the silver leopards; these elders have witnessed previous killings of livestock by the leopards. Only one case of a human kill by a silver leopard has been documented. As a response to the risks described above, the tribes have vowed to kill any silver leopard that crosses the preserve boundary, whether they pose an immediate threat or not.

In alliance with the tribes against the establishment of the preserve are the Oingo Boingo mining companies. Mining occurs in other areas of the country but has not yet been established in the proposed protection area as the mountainous terrain has made it difficult to access areas that are potentially mineral rich. However, recent developments in mining technology may make it possible to access these unproven areas. Mining currently accounts for 15% of Oingo Boingo's GDP and 5% of its employment. But those numbers may increase if mining is expanded into the proposed preserve, according to mining industry representatives. Creation of the preserve would prohibit any mining activity in this area into the foreseeable future.

EXERCISE

The time has come for the Oingo Boingo government to make a decision on whether or not to establish a wilderness preserve. The government has received a one-time $50 million cash offer from the Save Our Wildlife Federation as an incentive, and government delegates have met with representatives of the ecotourism and mining industries as well as the tribal elders to hear arguments for and against the preserve. The delegates now have to vote.

You are a delegate to the Oingo Boingo government who will vote on this issue. As a class, you will decide whether the preserve is established or not. The Oingo Boingo constitution requires a two-thirds majority to either support or prevent the establishment of the preserve. In the event that a two-thirds majority is not met, the issue is re-debated and votes are recast; this procedure is repeated until the two-thirds majority is reached to either support or deny the preserve.

PRE-CLASS PREPARATION:

1) Prior to class, spend time researching real-life conflicts in establishing wildlife preserves. Use that research as well as the background information above to write at least five pros and cons of establishing a wildlife preserve in Oingo Boingo to benefit silver leopards. You may find it helpful to organize your research by using a table similar to the one below:

Option	Pros	Cons
Establish the preserve		
Prevent the preserve's establishment		

IN-CLASS VOTE:

1) In class, discuss the list of pros and cons. Add any additional pros and cons to your personal list that others' have described.

2) After reviewing all of the pros and cons, cast your vote. Your instructor may have you complete an anonymous ballot or do an open vote by show of hands.

3) Was the two-thirds majority met? If so, discuss as a class why you voted the way you did. If not, open the floor to debate for 10–15 minutes. Both sides of the debate should be represented, and each participating speaker should be limited to a two-minute speech. Once the debate period has ended, complete a second vote. Was the two-thirds majority met this time? If so, why were some convinced to vote differently the second time than the first? If not, then repeat the debate-vote process described above until the two-thirds majority is met or class time runs out.

POST-CLASS REFLECTION:

1) Controversies related to the establishment of wildlife protection areas or preserves are often framed as a "people versus animals" conflict in which one side "loses" while the other side "wins." In your opinion, can wildlife and humans co-exist without one side losing? Can conflicts in wildlife protection areas or preserves be reduced or mitigated? If so, how? Use this particular case and your research of other wildlife preserve conflicts in supporting your argument of whether humans and wildlife can co-exist. Your response should be no longer than one double-spaced page.

Literature Cited:

Okech, R. N. 2010. "Wildlife-community Conflicts in Conservation Areas in Kenya." *African Journal on Conflict Resolution* 10:65–80.

Case VIIB

The Dingo that Divides Us: Controversies of Reintroducing and Protecting Predators

INTRODUCTION

Islands are particularly vulnerable to species extinctions. The biodiversity of animal communities on islands is unique because the biota is isolated from mainland populations and the species have evolved over time independently under different selective forces. However, islands are also highly vulnerable to species extinctions, especially when new animals are introduced to those islands. Native fauna have minimal defenses from competition and predation by these invaders. As a result, islands are often referred to as places of "concentrated extinctions" (Whittaker and Fernandez-Placios 2007).

Australia is not an exception to this rule. In fact, the island continent's ecosystem has been restructured over many thousands of years as a result of introductions of animals such as pigs, goats, cattle, camels, rabbits, buffalo, and deer as humans have colonized Australia (Bowman 2012). Approximately 4,000 years ago, the dingo (*Canis lupus dingo*) was introduced to Australia by seafarers from Asia and slowly became Australia's top predator as other predators such as the thylacine (*Thylacinus cynocephalus*) disappeared (Lester 2006; Johnson 2007). Interestingly, dingoes have contributed to few, if any, native animal extinctions since their introduction (Johnson 2007). Today, the dingo is classified as Australia's only native canid (QGDEHP 2013).

Since the arrival of Europeans to Australia's mainland over 200 years ago, introductions of new species and extinctions of native species have accelerated. For example, at least 18 species of Australian mammals have become extinct, which is half of the world total that went extinct during that same period; many other mammalian species are in serious decline (Bowman 2012). Much of this loss in biodiversity can be attributed to introductions of red fox (*Vulpes vulpes*) and feral cats. In addition, Europeans have a long history of shooting, poisoning, and fencing dingoes to protect grazing cattle and sheep. The world's longest fence was built to prohibit dingo migration. The fence stretches approximately 5,000 km from Jinbour in Queensland to the Eyre Peninsula; this fence is almost as long as the Great Wall of China (Lester 2006)! To the south of the fence, dingoes have been nearly extirpated and sheep are grazed. To the north of the fence, cattle and dingoes coexist but not without conflict (Lester 2006).

The fence provides a unique opportunity to study the interactions between dingoes and non-native predators such as red foxes and feral cats. Dingoes will hunt foxes and cats—sometimes for food but sometimes to reduce competition for shared prey resources (Lester 2006). Research has shown that red foxes are seven to 20 times more abundant south of the fence where dingoes are rare compared to north of the fence (Johnson and VanDerWal 2009). Further, native mammals are more abundant where dingoes are more plentiful compared to areas where they are rare or non-existent in certain sections north of the fence (Letnic et al. 2012).

In an effort to protect Australia's unique biodiversity and bring ecosystems back in balance, some biologists and ecologists are calling for the Australian government to reintroduce dingoes where they have been extirpated and to protect dingoes from poisoning or shooting where they are present. These ideas have been met with strong opposition from those directly or indirectly involved in the cattle and sheep industry. They contend that dingoes should NOT be considered to be native species and, as such, do not warrant government protection. The debate has divided public opinion.

EXERCISE

This case study is loosely based on the facts on current conflicts related to dingoes in Australia. To fully complete this case study, you will be required to prepare information prior to the class exercise and to complete a post-class activity.

PRE-CLASS PREPARATION:

Spend some time researching RELIABLE resources to answer following questions:

1) What defines a native species?

2) Should dingoes be considered native? Why or why not?

3) What impact do dingoes have on their environment?

4) What are the various ways in which humans and dingoes can conflict?

5) Are there ways to prevent conflicts between dingoes and humans that would be easy for humans to adopt on a widespread scale?

IN-CLASS EXERCISE:

Two major coalitions have formed around the issue of dingo reintroduction in Australia: "Citizens for Dingoes" and "Citizens against Dingoes." Each coalition is made up of various parties who are either directly affected or simply have an opinion on this issue. The coalitions are working to meet with the citizens of Australia to convince them to sign a petition that either supports or rejects dingo reintroduction (depending on the coalition's position). Each coalition is working to get more signatures than the other.

Your instructor will assign you to one of these two coalitions. Meet with your fellow group members, and identify THREE spokespeople to represent the various parties who may join your particular coalition. Assist those spokespeople in developing a three-minute persuasive argument that would convince the average Australian citizen to sign their petition. The spokesperson will deliver the argument to the class. As you prepare and listen to all of the arguments, make any notes that may supplement the research that you completed prior to class.

POST-CLASS EXERCISE:

Research a case where reintroducing or protecting a predator has created controversy in North America using RELIABLE resources. Summarize the case in a single paragraph. Then, answer the following questions:

1) How is your chosen case similar and different to the dingo issue in Australia? Describe two similarities and two differences.

2) Consider your own views on the controversy described in both the dingo reintroduction issue as well as your chosen case. Are your views on both controversies similar or different? Why or why not?

3) Why are some individuals, regardless of their country of origin, against reintroduction of predators while others support reintroduction? Describe three factors that may influence one's views on such issues.

Literature Cited:

Bowman, D. 2012. Bring elephants to Australia? *Nature* 482:30.

Johnson, C. N. 2007. Australia's Mammal Extinctions: A 50,000-Year History. Cambridge, UK: Cambridge University Press.

Johnson, C. N., and J. VanDerWal. 2009. "Evidence That Dingoes Limit Abundance of a Mesopredator in Eastern Australian Forests." *Journal of Applied Ecology* 46:641–646.

Lester, B. 2006. "The Dingo Divide." *Cosmos*. Accessed July 2013, http://www.cosmosmagazine.com/features/the-dingo-divide/.

Letnic, M., E. G. Ritchie, and C. R. Dickman. 2012. "Top Predators as Biodiversity Regulators: The Dingo *Canis lupus* Dingo as a Case Study." *Biological Reviews* 87:390–413.

Queensland Government Department of Environment and Heritage Protection (QGDEHP). 2013. Queensland Government Department of Environment and Heritage Protection. Accessed July 2013, http://www.ehp.qld.gov.au/wildlife/livingwith/dingoes/index.html.

Whittaker, R. J., and J. M. Fernández-Palacios. 2007. Island Biogeography: Ecology, Evolution, and Conservation, 2nd edition. Oxford, UK: Oxford University Press.

section 8

Freshwater and Marine Ecosystems and Fisheries Issues

Case VIIIA

Those Darn Dams! The Case of Gavins Point Dam, South Dakota-Nebraska

INTRODUCTION

State and federal natural resource managers and policy makers today are challenged with the task of balancing interests of different user groups. This case study is loosely based on a real on-going issue where state and federal agencies and policy makers on the Missouri River are working on a plan to manage Lewis and Clark Lake, a Missouri River reservoir formed by Gavins Point Dam.

The United States Army Corps of Engineers (USACOE) has constructed, and presently operates, six dams on the mainstem of the Missouri River; these six dams have created six reservoirs (see map of the Missouri River basin at http://www.cerc.usgs.gov/rss/visualize/visualize.htm). Water levels in the mainstem river and the reservoirs are controlled through dam operation for the purposes of flood control, irrigation, navigation, power, municipal water uses, wildlife, and recreation. The first dam (Fort Peck) was completed in 1937, and the last dam (Big Bend Dam) was completed in 1963. Lewis and Clark Lake was formed in 1955 by the completion of Gavins Point Dam.

One of the major issues with dams is their short lifespan due to increased sediment accumulation rates caused by reductions in water flow. Recent estimates indicate sediment accumulation rates on the Missouri River reservoirs average 89,700 acre-feet per year. It is estimated that this sedimentation results in a loss of water storage to supply more than 800,000 people with 100 gallons per day for an entire year. (Missouri Sedimentation Action Coalition. http://www.msaconline.com; Accessed: 8/1/13).

Where does all of this sediment end up? On Lewis and Clark Lake, it appears to "pile up" at the upper end of the reservoir at the confluence of the Niobrara River (see photos of this accumulation at http://www.msaconline.com). The accumulation of sediment in the reservoir has reduced its original water storage capacity of 492,000 acre-feet. Consequently, electricity production, municipal water, and rural irrigation use capabilities have also decreased. Additionally, thousands of recreational boaters and anglers from other towns far outside of Yankton and even from states other than South Dakota and Nebraska also visit the lake contributing tourist dollars to the local economy of Yankton and the surrounding area. The delta area is seen as a "navigational hindrance" to these recreational boaters and anglers and many have voiced their frustrations to local businesses. For these various reasons, citizens of Yankton and other nearby towns that rely on Lewis and Clark Lake for power, water, and tourist dollars would like to retain the use of the reservoir but would like to see the sediment removed from the delta area either through dredging or flushing.

However, the delta may provide some ecological benefits that have not yet been fully explored. Research from universities as well as South Dakota Game, Fish, and Parks and the Nebraska Game and Parks Commission indicates that the delta area serves as an important spawning and nursery habitat for many native and imperiled fishes such as sauger (*Sander canadensis*) and silver chub (*Macrhybopsis storeriana*) and may be important nesting habitat for endangered birds such as the interior least tern (*Sterna antillarum*) and the piping plover (*Charadrius melodus*). Wildlife conservation groups (i.e., private citizen advocacy groups who are not employed as agency biologists) contend that the research conducted by the universities and the two state agencies support preserving the delta as is and that nothing should be done to move the sediment.

Given that many operating licenses for many major dams are not being renewed and that many smaller dams have now been removed, many anti-dam organizations have begun fighting for removal of major dams, even on the Missouri River. These groups believe that even if the existing sediment is removed, there is nothing to prevent sediment from reestablishing in the Niobrara River delta. Further, removal of the dams may restore the ecological function of the river, which would aid in the recovery of the imperiled fish and birds currently using the delta area and would naturally remove sediment from the upper end of the reservoir. However, removal of the Gavins Point Dam may be more costly than removing sediment and would impact municipal and rural water use. Secondly, the dam may serve as a barrier to the upstream migration of many aquatic invasive species such as bighead and silver carp (*Hypophthalmichthys* spp.). To see other harmful negative effects carp can have, visit this link on YouTube: http://www.youtube.com/watch?v=6rF-STc3-Js (Accessed: 8/1/13).

The Governors of South Dakota and Nebraska are forming a "Delta Task Force." The task force will be composed of members of three stakeholder groups:

> citizens of Yankton and nearby towns,
> wildlife conservation groups; and
> anti-dam organizations.

The charge of the task force will be to make a recommendation to the Governors on potential solutions to the issues described above by exploring three options:

> remove the sediment from the Niobrara River delta by either dredging or flushing;
> remove Gavins Point Dam;
> allow the reservoir and the delta to remain as is (i.e., the "do nothing" option).

The Governors will approach the USACE with recommendations of the task force with the hopes of implementing the option desired.

EXERCISE

Your instructor may elect to complete this assignment as either an in-class discussion/debate activity or an individual out-of-class written assignment. Directions for both activities are provided.

IN-CLASS DISCUSSION/DEBATE DIRECTIONS:

1) Spend time prior to class researching the pros and cons of each option using reliable resources. For example, you might find other cases where dams have had a positive or negative impact on wildlife and fish, the monetary and environmental costs of dredging sediments, or the monetary and social costs of removing dams, just to name a few. Use your research and the background information above to write at least three pros and three cons for each option prior to class. You may find it helpful to organize your research by using a table similar to the one below:

Option	Pros	Cons
Remove the sediment		
Remove Gavins Point Dam		
Do nothing		

2) In class, your instructor will assign you to one of the stakeholder groups. Meet with your fellow group members to discuss the pros and cons of each option that you came up with prior to class. Compile a master list of pros and cons for the three options.

3) Once you have your master list of pros and cons, discuss which option your particular stakeholder group would support and which options your group would oppose. Work with your group to write a 3–5 minute position statement that advocates for your preferred option. In crafting your position statement, address why your chosen option is best and why the other options are less desirable using the pros and cons list that your group created in question #2 above. Elect a group spokesperson to deliver your position statement orally to the class.

4) Listen to the position statements of the other two groups and jot down any additional pros and cons that your group did not identify in question #2 above. Either at the end of class or before the next class period, write a one- or two-paragraph response to the following question: If you were either the Governor of South Dakota or Nebraska, which option would you support after hearing the various sides of the issue? Support your answer with the pros and cons of each option that you generated from your own list, your group discussion, and the position statements of all of the stakeholder groups you heard in class.

OUT-OF-CLASS INDIVIDUAL WRITTEN ASSIGNMENT DIRECTIONS:

1) Research the pros and cons of each option using reliable resources. For example, you might find other cases where dams have had a positive or negative impact on wildlife and fish, the monetary and environmental costs of dredging sediments, or the monetary and social costs of removing dams, just to name a few. Use your research and the background information above to write at least three pros and three cons for each option in three to six paragraphs.

2) In one to three paragraphs, describe which option each stakeholder group would support and which options would be opposed. Justify the position of each stakeholder group using the pros and cons list that you created in question #1 above.

3) In one or two paragraphs, address the following question: If you were either the Governor of South Dakota or Nebraska, which option would you support after examining the various sides of the issue? Support your answer with the pros and cons of each option that you generated in question #1 above.

Case VIIIB

Super Salmon! Should Genetically Engineered Salmon be Allowed on US Dinner Plates?

INTRODUCTION

A genetically-engineered (GE) organism is one whose genetic material has been altered by human intervention to produce a desirable trait or group of traits. Genetically-engineered crops were first introduced into the United States in 1996, producing traits such as disease or drought resistance and extended shelf lives. The presence of these GE crops in supermarkets, either as whole foods or food additives, is prevalent since the US Food and Drug Administration (FDA) deemed these food products safe for human consumption.

As of early 2013, no GE animal products have been approved for human consumption by the FDA. However, the first GE animal could be found on US dinner plates by the end of 2013 in the form of AquAdvantage™ salmon produced by AquaBounty Technologies. This salmon is the result of inserting a growth hormone regulating gene from a Pacific Chinook salmon (*Oncorhynchus tshawytscha*) and a growth gene promoter from an ocean pout (*Zoarces americanus*) to the 40,000 genes already found in an Atlantic salmon (*Salmo salar*). The Chinook salmon gene promotes faster growth rates to those currently observed in Atlantic salmon and the promoter from the ocean pout encourages growth to occur all year long rather than just during the warmer months of the year. The result, as indicated by AquaBounty Technologies, is an "improved" Atlantic salmon that takes only 16–18 months to reach market size as compared to the three years it currently takes for non-genetically-engineered Atlantic salmon to reach market size under current aquaculture practices. Reduced time to market correlates to less feed required to raise salmon in captivity and potential reduced costs to the consumer.

The possible approval of AquAdvantage™ salmon for human consumption has been controversial. In 2011, Congress halted the FDA process for approving AquAdvantage™ salmon, citing that not enough information and research yet exists in regards to potential impacts on human health or the risks that GE salmon may impose on native wild salmon should the captive GE salmon escape from their pens (which is a common issue in aquaculture facilities). Despite Congressional intervention, the FDA resumed the process to approve AquAdvantage™ salmon in late 2012 and it is speculated that the approval process will be completed by the end of 2013.

Letter writing campaigns and public meetings with the FDA are currently being organized by several stakeholder groups to either encourage or discourage FDA approval of AquAdvantage™ salmon for commercial sale and human consumption. These stakeholder groups include the following:

Support Approval	Oppose Approval
Commercial salmon aquaculturists	Commercial salmon fishermen
Restaurant owners associations	Consumer advocacy groups
	Environmental advocacy groups

In this case study, you will play the role of one or more of these groups and deliver a convincing argument to the FDA on whether they should approve or deny human consumption of AquAdvantage™ salmon. You will also explore and describe your own views on GE organisms.

EXERCISE

Your instructor may elect to complete this assignment as either an in-class "public meeting" forum or an individual out-of-class "letter writing" assignment. Directions for both activities are provided.

IN-CLASS PUBLIC MEETING FORUM DIRECTIONS:

1) Spend time prior to class researching the following topics as they relate to the controversy surrounding AquAdvantage™ salmon.

    - The process of raising AquAdvantage™ salmon

    - Environmental issues related to the aquaculture practices of marine salmon

    - Moral and ethical concerns of consuming GE crops and animals

    - Human health concerns of GE food consumption

    - US labeling requirements for GE crops and animals

    - Economic impacts of GE foods, aquaculture, commercial fishing

2) Use your research above to describe the arguments of the various stakeholder groups in regards to their support or opposition to AquAdvantage™ salmon. Your descriptions should focus on various factors such as the economic, ethical, environmental, and human health impacts these salmon may have, either positively or negatively. Note that some groups may focus on one impact to the exclusion of others, but some groups may argue their case based on several impacts. You may find it helpful to organize your research by using a table similar to the one below:

Stakeholders	Potential Impacts			
	Economic	Environmental	Ethics	Human Health
Commercial salmon aquaculturists				
Commercial salmon fishermen				
Consumer advocacy groups				
Environmental advocacy groups				
Restaurant owners associations				

3) In class, your instructor will assign you to one of the five stakeholder groups. Meet with your fellow group members to discuss the points of that particular stakeholder group that you came up with prior to class. Compile a master list of those points. Once you have your master list, work

with your group to write a 3–5 minute position statement that advocates for either support or opposition (depending on your assigned group). Elect a group spokesperson to deliver your position statement orally to the class.

4) Listen to the position statements of the other stakeholder groups and jot down any additional arguments that you did not come up with for question #2 above. Either at the end of class or before the next class period, write a one- or two-paragraph response to the following questions:

- Do you believe that the FDA should support or deny AquAdvantage™ salmon for commercial sale and human consumption? Support your answer with the research and arguments that were generated for questions #1–3 above.

- Do you believe that GE crops and animals should have required labeling to identify the food products as GE? Why or why not?

OUT-OF-CLASS LETTER WRITING ASSIGNMENT:

1) Research the following topics as they relate to the controversy surrounding AquAdvantage™ salmon.

- The process of raising AquAdvantage™ salmon

- Environmental issues related to the aquaculture practices of marine salmon

- Moral and ethical concerns of consuming GE crops and animals

- Human health concerns of GE food consumption

- US labeling requirements for GE crops and animals

- Economic impacts of GE foods, aquaculture, and commercial fishing

Use your research above to describe the arguments of the various stakeholder groups in regards to their support or opposition to AquAdvantage™ salmon. Your descriptions should focus on various factors such as the economic, ethical, environmental, and human health impacts these salmon may have, either positively or negatively. Note that some groups may focus on one impact to the exclusion of others, but some groups may argue their case based on several impacts. You may find it helpful to organize your research by using a table similar to the one below:

Stakeholders	Potential Impacts			
	Economic	Environmental	Ethics	Human Health
Commercial salmon aquaculturists				
Commercial salmon fishermen				
Consumer advocacy groups				
Environmental advocacy groups				
Restaurant owners associations				

2) Write a letter to the FDA in support or opposition of approval of AquAdvantage™ salmon for commercial sale and human consumption from the perspective of each of the five stakeholder groups (i.e., five letters total). Support your stakeholder group's position with the research you described for question #1 above. Remember, the FDA will be reading hundreds (perhaps thousands) of letters. Your letter should be concise but impactful. Keep your letter no longer than one double spaced page.

3) Now that you understand the viewpoints of the various stakeholder groups, write a one- or two-paragraph response to the following questions:

- Do you believe that the FDA should support or deny AquAdvantage™ salmon for commercial sale and human consumption? Support your answer with the research and arguments that were generated for questions #1–3 above.

- Do you believe that GE crops and animals should have required labeling to identify the food products as GE? Why or why not?